THE POETRTY OF
PALLADIUM

The Poetrty of Palladium

Walter the Educator

SKB

Silent King Books a WhichHead Imprint

Copyright © 2023 by Walter the Educator

All rights reserved. No part of this book may be reproduced in any manner whatsoever without written permission except in the case of brief quotations embodied in critical articles and reviews.

First Printing, 2023

Disclaimer
This book is a literary work; poems are not about specific persons, locations, situations, and/or circumstances unless mentioned in a historical context. This book is for entertainment and informational purposes only. The author and publisher offer this information without warranties expressed or implied. No matter the grounds, neither the author nor the publisher will be accountable for any losses, injuries, or other damages caused by the reader's use of this book. The use of this book acknowledges an understanding and acceptance of this disclaimer.

"Earning a degree in chemistry changed my life!"
- Walter the Educator

dedicated to all the chemistry lovers, like myself, across the world

CONTENTS

Dedication v

Why I Created This Book? 1

One - Element Of Grace 2

Two - Oh, Palladium 4

Three - Science's Hall Of Fame 6

Four - Power Of Palladium 8

Five - Shimmering Presence 10

Six - Element Divine 12

Seven - Cherished And Rare 14

Eight - Science And Art 16

Nine - Captured Our Heart 18

Ten - Guardian Of Air 20

Eleven - Future Secure 22

Twelve - Shine So Bright 24

Thirteen - Promise Unfurled 26

Fourteen - Through The Night 28

Fifteen - Ignite The Fire 30

Sixteen - Palladium Gleams 32

Seventeen - Mender Of Bones 34

Eighteen - Propelling Us Higher 36

Nineteen - Innovation And Dreams 38

Twenty - Empowering Progress 40

Twenty-One - Hearts And Minds 42

Twenty-Two - Hope Restarts 44

Twenty-Three - Harmony Refined 46

Twenty-Four - Form And Break 48

Twenty-Five - Gift To Humankind 50

Twenty-Six - For All Our Days 52

Twenty-Seven - Illuminated And Stark 54

Twenty-Eight - Precious Element 56

Twenty-Nine - Catalyst Of Health 58

Thirty - Secrets Reside 60

Thirty-One - Palladium's Presence 62

Thirty-Two - Change And Hope 64

Thirty-Three - Bright And New 66

Thirty-Four - Harbinger Of Change 68

Thirty-Five - Palladium, The Healer 70

About The Author 72

WHY I CREATED THIS BOOK?

Creating a poetry book about the chemical element of Palladium was an intriguing and unique endeavor. Palladium, a rare and lustrous metal, possesses various symbolic qualities that can inspire poetic exploration. This book delves into the elemental nature of Palladium, its properties, history, and applications, and intertwine them with themes of beauty, rarity, transformation, and resilience. By fusing science and art, this book can illuminate the hidden connections between the natural world and human emotions, inviting readers on a journey that combines scientific curiosity with the power of poetic expression.

ONE

ELEMENT OF GRACE

In the realm of elements, rare and bright,
Lies Palladium, a metal of sheer delight.
With a silvery sheen, it captures the eye,
A treasure of the Earth, soaring high.

Born in the depths, where nature's secrets lie,
Palladium emerges, shining in the sky.
Its atomic number, forty-six it bears,
A symbol of strength, resilient and rare.

A catalyst of change, it plays its part,
In chemical reactions, a work of art.
From catalytic converters to jewelry refined,
Palladium's versatility, a precious find.

Its atomic structure, a marvel of perfection,
With electrons dancing in harmonious connection.

Strong yet malleable, it bends to our will,
A symbol of resilience, a story to fulfill.
  In the world of science, it takes center stage,
With applications vast, spanning each age.
From medicine to technology, it lends a hand,
Palladium, a catalyst, forever grand.
  Oh, Palladium, element of grace,
You illuminate our world, in every place.
A shining testament to nature's design,
Forever we cherish, this treasure of mine.

TWO

OH, PALLADIUM

In the realm of elements rare and bright,
There exists a metal with silvery light.
Palladium, its name, a symbol of might,
A beacon of strength, shining through the night.
    Atomic number forty-six, it claims,
Protons and neutrons in its atomic frame.
With electrons dancing, a celestial ballet,
Palladium's essence, forever in play.
    Versatile in nature, it finds its place,
Catalytic converters, with elegance and grace.
In jewelry, it gleams, a precious embrace,
Palladium's allure, none can efface.
    Its atomic structure, a marvel untold,
A lattice of atoms, strong and bold.
A catalyst it becomes, in reactions untold,
Facilitating transformations, a story unfold.

In science, medicine, technology's embrace,
Palladium shines, leaving a lasting trace.
In laboratories, experiments take flight,
With Palladium's guidance, illuminating the night.

Oh, Palladium, element of grace,
Your brilliance illuminates our world's space.
A symbol of resilience, strength, and might,
In your presence, darkness takes flight.

Let us celebrate this metal rare,
A testament to human curiosity and care.
Palladium, we admire your gleaming soul,
Forever in our hearts, your story, forever told.

THREE

SCIENCE'S HALL OF FAME

In the realm of elements, Palladium shines so bright,
A metal of wonders, a captivating light.
With fifty-six protons, its atomic embrace,
A symphony of particles, a delicate grace.

Beneath its silver sheen, a secret unfolds,
A catalyst of change, a story yet untold.
In the depths of its core, a power resides,
Transforming molecules with precise tides.

From automobile engines to the artist's brush,
Palladium's versatility, an industry's hush.
Catalytic converters, purifying the air,
Reducing toxic emissions, showing it cares.

In jewelry it sparkles, a dazzling array,
A precious adornment, a radiant display.

A symbol of elegance, a treasure untold,
Palladium's allure, a sight to behold.

A metal strong and sturdy, with resilience untamed,
A companion in progress, its name ever acclaimed.
From laboratories to skyscrapers so tall,
Palladium stands firm, steadfast through it all.

So let us praise this element, noble and rare,
For its beauty, strength, and the wonders it shares.
Palladium, a gift to the world, we acclaim,
Forever glowing, in science's hall of fame.

FOUR

POWER OF PALLADIUM

In the realm of elements, a noble star,
A shining metal, Palladium by far.
With atomic number forty-six it reigns,
Among the periodic table's splendid domains.

A catalyst, it dances in the reaction's core,
Igniting sparks that were hidden before.
A silent force, it speeds up the pace,
Transforming molecules with elegant grace.

In the chambers of engines, it finds its place,
Catalytic converters, its noble embrace.
Transforming toxic fumes with each breath,
Reducing emissions, protecting life's breath.

Oh, Palladium, a jeweler's delight,
In lustrous creations, you shimmer so bright.

Adorned on fingers and wrists, a gleaming treasure,
Reflecting love's stories, a symbol of pleasure.
 In industries vast, your presence is profound,
From electronics to dentistry, you're renowned.
Your strength and versatility, a marvel to see,
An element of wonder, forever it will be.
 Oh, Palladium, with your atomic might,
A catalyst, a guardian, shining with light.
In science and industry, you reign supreme,
A precious metal, a scientist's dream.
 So let us celebrate, with a heartfelt cheer,
The beauty and power of Palladium, so clear.
For in its presence, we find hope and grace,
A shining star in the vast cosmic space.

FIVE

SHIMMERING PRESENCE

In the realm of elements, Palladium shines bright,
A radiant metal, a captivating light.
Its beauty, unmatched, in hues of silver and gray,
A precious jewel, where elegance holds sway.

A catalyst of change, it weaves a magical spell,
In industries diverse, where transformations dwell.
From automotive engines to the world of medicine,
Palladium's touch brings progress and discipline.

In catalytic converters, it transforms toxic fumes,
Reducing emissions, cleansing the earthly tombs.
A guardian of air, it breathes life into the sky,
A silent hero, unseen, but never shy.

In jewelry and art, it adorns with grace,
A symbol of elegance, a delicate embrace.

With strength and resilience, it withstands the test,
Enduring the ages, forever at its best.

From the depths of the Earth, it emerges with might,
A rare gift from nature, a beacon of light.
Palladium, a marvel, that captures our gaze,
A testament to science, in countless ways.

So let us celebrate, this element so grand,
Palladium, the jewel of the elemental band.
With its shimmering presence, it never fails,
To inspire and awe, as its story unveils.

SIX

ELEMENT DIVINE

In the realm of elements, a luminary gleams,
Palladium, the star that reigns supreme.
With atomic number forty-six, it stands,
A jewel of transition, in nature's hands.

A noble metal, elegant and rare,
Palladium shines with an aura so fair.
With silvery-white grace, it adorns the earth,
Unveiling its secrets, revealing its worth.

In catalysts, its prowess is revealed,
A catalyst supreme, to progress it wields.
From automotive engines, it lends a hand,
Transforming gases, purifying the land.

Its strength and resilience, a marvel to behold,
A shield against corrosion, it's unyielding and bold.
In jewelry and art, it finds its muse,
Crafting masterpieces, it never fails to enthuse.

    From the depths of mines to the heavens above,
Palladium emerges, a symbol of love.
Through science and industry, it paves the way,
A beacon of progress, a guide for each day.
    So let us celebrate this element divine,
A testament to beauty, strength, and shine.
Palladium, a treasure in nature's domain,
A promise of hope, forever it will remain.

SEVEN

CHERISHED AND RARE

In the realm of elements, a shimmering star,
Palladium, thou art, resplendent and far.
With elegance and grace, thy atoms align,
A noble metal, rare and divine.
　　Within thy core, protons dance in delight,
Electrons embrace, creating a sight,
Atomic number, forty-six, so true,
A symbol of strength, in all that you do.
　　Oh, Palladium, catalyst supreme,
In the realm of science, you reign supreme.
With a touch so gentle, you spur reactions,
Transforming molecules with your magic attractions.
　　In medicine's embrace, you heal and restore,
A guardian of health, forevermore.

Implants and devices, with your aid,
Enhancing lives, a debt never repaid.

In technology's realm, you shine so bright,
A conductor of dreams, a beacon of light.
Circuit boards sing with your electric flow,
Empowering progress, as innovations grow.

Jewelry adorned, with your lustrous hue,
Admired and treasured, forever anew.
A symbol of love, a token of grace,
Palladium, your beauty, none can replace.

Oh, Palladium, pure and strong,
A guardian of air, against pollution's wrong.
Cleansing the world, with your noble might,
Reducing emissions, a planet's delight.

Unyielding to time, corrosion's foe,
Palladium, thy resilience, all shall know.
A precious element, cherished and rare,
A gift from nature, beyond compare.

So, let us celebrate Palladium's might,
A dazzling element, shining so bright.
In science, medicine, and jewelry's embrace,
Palladium, forever, you hold a special place.

EIGHT

SCIENCE AND ART

In the depths of science, a treasure resides,
A metal of wonder, where brilliance abides.
Palladium, the catalyst, mighty and true,
Unlocks the potential, in all that we do.

In factories and labs, its power is known,
A guardian of progress, it proudly is shown.
Transforming reactions, with precision and grace,
Palladium's touch, a catalyst's embrace.

In jewelry it gleams, a symbol so rare,
Adorning the fingers of those who dare.
Its lustrous allure, a sight to behold,
A testament to elegance, wrapped in gold.

But art too finds solace, in Palladium's gleam,
Crafting masterpieces, like a vivid dream.
Sculpting its essence, with passion and might,
Creating beauty, that shines in the light.

And as the air thickens, with toxins and haze,
Palladium steps forth, a savior ablaze.
With powers untamed, it purifies the air,
Protecting our planet, with utmost care.

Oh, Palladium, your strength knows no bounds,
A guardian of progress, in science profound.
From industries vast, to jewelry so fine,
You shine through it all, a luminary divine.

Let us raise our voices, in a reverent cheer,
For Palladium, the element we hold dear.
A symbol of resilience, beauty, and more,
In science and art, forever you'll endure.

NINE

CAPTURED OUR HEART

In the realm of atoms, an element divine,
Palladium, a jewel in nature's design.
A guardian of air, with powers profound,
Cleansing our world, where purity is found.

With grace it captures, as the winds do blow,
The pollutants that tarnish, the toxins that flow.
A silent warrior, fighting emissions' plight,
Palladium's presence, a beacon of light.

In the realm of art, its beauty does gleam,
Adorning our bodies, a precious dream.
A metal so strong, yet delicate in form,
Palladium, a treasure, in jewelry to adorn.

Versatile and rare, it weaves a tale,
Of strength and elegance, a love that won't fail.

In labs and industries, it finds its place,
A catalyst of progress, with boundless grace.

From engines to electronics, it lends its might,
A catalyst for change, igniting innovation's flight.
Resilient and strong, in the face of corrosion,
Palladium, a symbol of endurance and devotion.

In medicine and science, it plays its part,
A healer of ailments, a catalyst of art.
A guardian of progress, with every breath,
Palladium, the element that conquers death.

Oh, Palladium, your presence profound,
A guardian of air, a treasure to be found.
In air and earth, in science and art,
With grace and strength, you've captured our heart.

TEN

GUARDIAN OF AIR

In the realm where air dances, a guardian stands,
A palladium presence, steadfast and grand.
With strength and resilience, it weaves its embrace,
A noble protector in this celestial space.

From dawn's first light to twilight's descent,
Palladium's essence, a heavenly scent.
A metal, ethereal, with secrets untold,
A marvel of nature, a treasure to behold.

In industries vast, its worth is renowned,
A catalyst for progress, innovation unbound.
From cars that glide silently on roads,
To jewelry that adorns and beauty bestows.

But beyond the realm of commerce and trade,
Palladium's true power begins to cascade.

It purifies air, an alchemist's touch,
Protecting the planet, we love so much.
    In art and in science, it finds its domain,
A muse for creators, a healer of pain.
For in a canvas stroke or a surgeon's hand,
Palladium's magic, a masterpiece planned.
    Enduring through time, with grace it prevails,
A symbol of hope, when all else fails.
Oh, palladium, guardian of air,
In your presence, we find solace and care.

ELEVEN

FUTURE SECURE

In the depths of Earth's embrace, a treasure lies untold,
A shimmering element of grace, a story yet unfold.
Palladium, the catalyst, in its noble form,
Unveiling secrets of the past and ushering a new dawn.

    Adorned by artisans, it gleams in silver's hue,
A noble metal, rare and pure, with wonders to pursue.
Crafted into jewelry, a symbol of enduring love,
Its luster captivates the heart, like stars that shine above.

    But beyond its beauty, a purpose deeply sewn,
Palladium, the guardian, a savior yet unknown.
In engines of progress, it toils without a pause,
Cleansing toxic fumes, the air it gently draws.

A silent hero in our cars, it tames the fiery spark,
Catalyzing transformation, igniting life's own arc.
From noble gases it purifies, a breath of freshened air,
A testament to its resilience, its strength beyond compare.

In labs and studios, where science finds its voice,
Palladium, the muse, inspires minds to rejoice.
A catalyst of innovation, it sparks the artist's dream,
Creating masterpieces, where beauty and science gleam.

Oh, Palladium, you bring solace and care,
With each molecule's embrace, hope fills the air.
In your presence, we find strength to endure,
A symbol of beauty, progress, and a future secure.

TWELVE

SHINE SO BRIGHT

In gilded realm of elements, behold Palladium's gleam,
A noble metal's grace, a poet's vivid dream.
Amidst the periodic table's grand symphony,
Palladium sings a tale of purity and harmony.

A guardian of the air, it purifies with might,
Cleansing our world, dispelling darkness, bringing light.
Through catalytic prowess, it orchestrates a dance,
Enabling progress, innovation's grand advance.

In engines' fiery hearts, it plays a vital role,
A silent guardian, protecting with its soul.
A catalyst of strength, a shield against decay,
Palladium's resilience never fades away.

From jewelry's delicate touch to scientific quest,
Palladium's allure, a brilliance manifest.

Its lustrous beauty, beyond compare,
Inspires art and science, a timeless affair.

In cars and electronics, its presence never fades,
A silent force propelling our technological crusades.
In medicine's embrace, it heals with gentle grace,
A precious element, imbued with life's embrace.

Oh, Palladium, your worth cannot be measured,
A treasure of the Earth, a gem forever treasured.
With every breath we take, your essence we inhale,
A reminder of your power, a story we regale.

In the tapestry of elements, you shine so bright,
Palladium, our guardian, our guiding light.

THIRTEEN

PROMISE UNFURLED

In the realm of progress, a catalyst's might,
Palladium shines, a beacon of light.
Its noble demeanor, a symbol of grace,
An element rare, in the periodic place.

A guardian, it stands, amidst chemical strife,
Unyielding and strong, it shapes a new life.
With platinum's kin, it shares its embrace,
A metal of promise, it leaves no trace.

In engines it dwells, with a purpose profound,
Igniting the flames, as the wheels spin around.
A catalyst true, it cleanses the air,
A guardian unseen, in the atmosphere.

In labs it's revered, a scientist's dream,
A catalyst supreme, where wonders gleam.

In art and science, it weaves a grand tale,
Unleashing the genius that will never fail.
   A symbol of hope, in a world so vast,
Palladium endures, its beauty will last.
For in its allure, a future secure,
A world transformed, where dreams can endure.
   So let us embrace this element rare,
A catalyst's gift, beyond compare.
For in its presence, a promise unfurled,
Palladium, the catalyst for a better world.

FOURTEEN

THROUGH THE NIGHT

In the realm of atoms, a metal divine,
Palladium, a catalyst, I proudly define.
A conductor of change, an alchemist's dream,
Unlocking the secrets of life's intricate scheme.

In the heart of innovation, you do reside,
Igniting progress with each step beside.
As a catalyst, you accelerate the course,
Transforming the ordinary into a mystical force.

Palladium, you purify, cleanse, and refine,
Removing impurities, a touch so divine.
In the crucible of science, you dance with precision,
Enabling breakthroughs, a catalyst's mission.

A guardian of art, you inspire creation,
A painter's brushstroke, a sculptor's elation.

With elegance and grace, you lend your hand,
Crafting masterpieces that forever expand.

Endurance, resilience, your essence profound,
A symbol of strength, forever renowned.
Through trials and tribulations, you endure,
A testament to the spirit that is pure.

Palladium, your versatility knows no bounds,
From cars to jewelry, your presence resounds.
An ally to nature, you shine without shame,
A guardian of Earth, forever our aim.

Oh, Palladium, your beauty, a beacon of light,
A symbol of hope, guiding us through the night.
In the vast universe of elements, you stand tall,
Palladium, a catalyst, for one and all.

FIFTEEN

IGNITE THE FIRE

In the realm of elements, a shining star,
Palladium, catalyst, you've come so far.
A metal rare, with a heart so pure,
Your presence, a catalyst, shall endure.

    In labs and factories, where progress thrives,
You aid the reactions, where science strives.
Your touch, a magic, that speeds up the dance,
Unleashing potential, with every chance.

    Through purification, you cleanse the air,
Binding with noble gases, removing despair.
A guardian of purity, you take their place,
Transforming the elements, with elegance and grace.

    In the realm of art, you inspire the soul,
A muse for creation, making us whole.

From brushstrokes to melodies, you ignite the fire,
Enveloping the world, with creative desire.

In technology's grasp, you find your home,
Powering devices, where innovation roams.
From circuitry to fuel cells, you pave the way,
Empowering progress, each and every day.

Oh, Palladium, you're a beacon of light,
Enduring and resilient, shining so bright.
A symbol of hope, in a world that's in need,
Your presence, a catalyst, for a better world, indeed.

SIXTEEN

PALLADIUM GLEAMS

In the realm of elements, noble and rare,
There lies a metal, beyond compare.
Palladium, a name that echoes with grace,
A catalyst of progress, in every space.

It dances with atoms, in a cosmic ballet,
Igniting innovation, along the way.
A catalyst of change, it sparks the flame,
Transforming the ordinary, to the extraordinary, with no shame.

In the depths of laboratories, it does reside,
Unleashing its power, with passion and pride.
A purifier of noble gases, it claims its might,
Cleansing the air, like a beacon of light.

But beyond the confines of science and labs,
Palladium inspires art, where beauty collabs.

In brushstrokes and melodies, it finds its voice,
Creating masterpieces, that make hearts rejoice.

In the heart of engines, it finds its place,
Powering cars, with a steady embrace.
Electronics hum with its radiant hues,
Connecting the world, with each vibrant fuse.

And in the realm of medicine, Palladium gleams,
Healing the wounded, mending broken dreams.
A catalyst of hope, in the hands of healers,
Restoring bodies and souls, like mystical sealers.

Oh, Palladium, your beauty knows no bounds,
From the depths of the Earth, to the stars that astound.
A catalyst of dreams, in every endeavor,
Guiding us forward, to a brighter forever.

SEVENTEEN

MENDER OF BONES

In the realm of elements, a shimmering star,
Palladium, thy name, revered from afar.
A metal of wonder, of secrets untold,
A symphony in atoms, a treasure to behold.

In art's embrace, thy beauty thrives,
A palette of dreams where creativity thrives.
Brushstrokes of inspiration, colors divine,
Palladium's essence, a muse so fine.

In engines and circuits, thy power unleashed,
A conductor of energy, a force released.
Revving hearts of steel, igniting the soul,
Palladium's presence, making machines whole.

In the halls of medicine, thy touch heals,
A balm for wounds, a solace that appeals.

Healer of hearts, mender of bones,
Palladium's potency, in remedies we've known.
    Versatile and enduring, thy spirit soars,
A beacon of hope, progress it explores.
From canvas to engine, through science's gate,
Palladium, forever, our futures await.
    So let us celebrate, this element rare,
Palladium, a gem beyond compare.
In art, technology, and healing's embrace,
Palladium, forever, a symbol of grace.

EIGHTEEN

PROPELLING US HIGHER

In the realm of elements, a gem beyond compare,
There lies Palladium, with grace so rare.
A catalyst it is, in art and in technology,
A muse for innovation, a creator's key.

    In vibrant hues, it adorns the artist's brush,
Breathing life into canvases, a master's touch.
A conductor of dreams, it sparks the mind,
Unleashing creativity, one of a kind.

    Through circuits and wires, it hums with power,
Igniting engines, propelling us higher.
Electronics dance to its shimmering beat,
A symphony of progress, a rhythm so sweet.

    But beyond the realm of human design,
Palladium's healing touch begins to shine.

In medicine's hands, it mends the broken,
A soothing balm for the wounded, a promise unspoken.
Versatile and enduring, it knows no bounds,
A beacon of hope, where solace is found.
It guides us forward, towards a brighter day,
A catalyst for progress, lighting the way.
Oh Palladium, your secrets unfold,
As you power machines and heal the soul.
A gem of grace, forever it will be,
In art, technology, and healing, a symphony.

NINETEEN

INNOVATION AND DREAMS

In the realm of elements, a marvel stands,
A noble metal, Palladium by name.
With grace and poise, it takes commanding hands,
To wield its power, to cherish its flame.

A catalyst, it sparks a wondrous dance,
Unleashing change with every tiny spark.
In chemistry's domain, it takes a chance,
Transforming molecules, leaving a mark.

In art's ethereal realm, it finds its place,
A muse for sculptors, painters, and creators.
Its lustrous shine, a canvas to embrace,
Reflecting beauty, capturing sensations.

In technology's embrace, it finds its role,
A conductor of innovation and dreams.

From circuits to fuel cells, it takes its toll,
Empowering progress, pushing the extremes.
    In medicine's domain, it heals and mends,
A trusted ally, a savior to behold.
With healing hands, its impact transcends,
Reviving hope, as stories are retold.
    Oh, Palladium, a beacon of light,
Guiding humanity towards a brighter sphere.
With every discovery, you ignite,
Insight and progress, forever near.
    So let us celebrate this noble metal,
A symbol of endurance, strength, and grace.
Embrace its power, let its presence settle,
And pave the way to a promising space.

TWENTY

EMPOWERING PROGRESS

In the realm of art, a gleaming hue,
Palladium, thou art forever true.
A catalyst for creativity's bloom,
A muse for painters, a poet's room.

Brush strokes dance upon the canvas pure,
Capturing light, emotions demure.
A metal rare, yet in art it finds,
A place to shimmer, to eternally bind.

In technology's realm, a silent force,
Palladium, thou art a source.
Cogs and gears, machines in motion,
Harnessing power, igniting devotion.

Engines roar with fiery might,
Palladium's touch, an engine's flight.

Revolutionizing, propelling us forth,
Advancing humanity, from south to north.

In the realm of healing, a gentle touch,
Palladium, thou art needed much.
A guardian of health, a doctor's aid,
In medicines potent, you are laid.

Healing wounds, curing ailments deep,
Palladium's power, a promise to keep.
Aiding humanity in times of despair,
With every heartbeat, showing you care.

Palladium, a metal of wonders untold,
Guiding us forward, with a story bold.
In art, in tech, in medicines too,
Empowering progress, reviving hope anew.

TWENTY-ONE

HEARTS AND MINDS

In the heart of Earth's embrace, a treasure lies,
A metal rare, with grace that mesmerizes.
Palladium, thy essence, a celestial dance,
A symphony of atoms, in elegance enhance.

A catalyst of dreams, you spark creations bright,
From art's vibrant palette to technology's light.
Inspiring innovation, with every touch,
Your presence ignites progress, oh, so much.

In engines you reside, a force to propel,
Powering the vessels that traverse the swell.
With strength and endurance, you lead the way,
Guiding humanity, towards a brighter day.

In medicine's realm, you heal with gentle might,
A guardian of health, a beacon of light.

From ailments you liberate, with tender care,
A soothing touch, relief beyond compare.
 Palladium, thy beauty, a shimmering embrace,
A metal that defies time, with elegance and grace.
In every noble heart, your essence blooms,
A symbol of resilience, dispelling gloom.
 Oh, Palladium, thy legacy shall endure,
A testament to human ingenuity, pure.
Forever you shall shine, in our hearts and minds,
A symbol of progress, that eternally binds.

TWENTY-TWO

HOPE RESTARTS

In the depths of Earth's embrace, hidden and rare,
Lies a precious metal, beyond compare.
Its name is Palladium, a force untamed,
Igniting progress, where dreams are reclaimed.

With strength and grace, it powers the machines,
In engines it roars, a symphony of dreams.
From cars to planes, it propels us forward,
A catalyst of change, relentless and onward.

In circuits and wires, it pulses and flows,
Electronics come alive, a vibrant glow.
Palladium dances, conducting the sound,
A symphony of technology, profound.

But beyond the realms of mechanics and gears,
Palladium whispers, and healing appears.

In medicine's hands, it mends broken hearts,
A soothing touch, where hope restarts.

From shattered souls to mending bones,
Palladium's grace in hospitals is known.
A beacon of hope, it shines through the pain,
Guiding us towards healing, again and again.

Oh, Palladium, element so divine,
Your versatility, a treasure to find.
In engines, electronics, and healing's embrace,
You weave a tapestry of progress and grace.

So let us celebrate, this metal rare,
Palladium's legacy, beyond compare.
A symbol of strength, innovation, and light,
Guiding humanity, towards a future so bright.

TWENTY-THREE

HARMONY REFINED

In gleaming grace, Palladium does shine,
A metal rare, with secrets so divine.
A catalyst of progress, it holds the key,
Unleashing wonders for humanity to see.

In circuits and wires, it finds its place,
A conductor of dreams, a tech-savvy embrace.
Electronics hum with its guiding might,
Palladium's touch, igniting innovation's light.

Beyond the screens that captivate our gaze,
Palladium whispers of healing ways.
In medicine's realm, it takes its stand,
A soothing balm, a gentle, helping hand.

From engines roaring with power and might,
To the hearts that beat, embracing life's flight,
Palladium's presence, a force so strong,
Ignites the fires that push us along.

A symbol of endurance, it does persist,
Through trials and challenges, it won't desist.
Palladium, a beacon of hope untold,
Guiding us towards a future yet unfold.

So let us cherish this noble metal's might,
And marvel at its wonders, shining bright.
For in Palladium's essence, we shall find,
A world of progress, harmony refined.

TWENTY-FOUR

FORM AND BREAK

In the realm of technology, you shine bright,
Palladium, a metal of remarkable might.
With electrons dancing, conducting the flow,
You power the devices that we all know.

From circuit boards to smartphones so sleek,
You lend your strength, never weak.
In wires and connectors, you hold steady,
Enabling communication, always ready.

But you're not confined to the realm of machines,
In medicine, your wonders are seen.
As a catalyst, you aid in healing,
Through palladium-based drugs, revealing.

Your touch brings hope to those in pain,
A soothing balm, a gentle rain.
With every bond you form and break,
You mend the broken, for their sake.

Palladium, you're a beacon of progress,
An element that never ceases to impress.
Innovation flourishes under your gaze,
Igniting our minds, setting ablaze.

So let us salute your noble presence,
A symbol of resilience and essence.
In technology and medicine, you're adored,
Palladium, forever our guiding chord.

TWENTY-FIVE

GIFT TO HUMANKIND

In the realm of elements, one shines bright,
Palladium, a conductor of light.
A metal rare, with a silvery sheen,
Unveiling secrets yet to be seen.

    In circuits and wires, it finds its place,
Guiding electrons with seamless grace.
Technology's ally, it paves the way,
Innovation's companion, day by day.

    In medicine's realm, it offers a hand,
Healing wounds, a noble command.
A guardian of health, with prowess untold,
Restoring life, in ways yet unfold.

    Its endurance unmatched, a symbol of might,
Enduring trials, shining through the night.

A steadfast companion, in times of strife,
Igniting hope, breathing renewed life.
    Oh, Palladium, catalyst of dreams,
A beacon of progress, it always seems.
In laboratories, where wonders take flight,
It dances with elements, gleaming so bright.
    So let us celebrate this element grand,
With gratitude, we reach out our hand.
For Palladium, a gift to humankind,
A key to a future, we long to find.

TWENTY-SIX

FOR ALL OUR DAYS

In the realm of healing, a presence strong,
A shimmering light, a soothing song,
Palladium, the element divine,
A gift to humanity, a treasure to find.

Within its core, a power untold,
A catalyst of change, a substance bold,
Mending broken hearts, a mender's touch,
A remedy for ailments that hurt so much.

In the hands of healers, it takes its form,
A symbol of hope, a shield in the storm,
With gentle grace, it whispers its plea,
"Embrace my essence, let me set you free."

Through veins and vessels, it gracefully flows,
Revitalizing life, where hope once froze,
A beacon of strength, a guide in the night,
Palladium, the healer, shining so bright.

In laboratories, it dances and gleams,
Unlocking secrets, unraveling dreams,
A catalyst for progress, a scientific delight,
Palladium, the innovator, soaring to new heights.

From microchips to fuel cells, it leads the way,
Pushing boundaries, with each passing day,
A symphony of electrons, a conductor of change,
Palladium, the pioneer, never to estrange.

Oh, Palladium, you bring us healing and light,
A shimmering star, shining so bright,
In your presence, we find solace and grace,
A testament to your enduring embrace.

From medicine to science, you pave the way,
Guiding humanity towards a brighter day,
Oh, Palladium, we sing your praise,
A symbol of progress, for all our days.

TWENTY-SEVEN

ILLUMINATED AND STARK

In the realm of science, a gleaming treasure lies,
A catalyst of hope, that in our hearts resides.
Palladium, a shimmering jewel, so precious and rare,
A metal that whispers tales of healing in the air.

In medicine's embrace, it finds its noble place,
A guardian of health, a savior in the chase.
With gentle touch, it weaves a tapestry of grace,
Mending broken souls, bringing solace to each face.

Its versatility, a symphony of endless might,
Unfolding miracles, like stars in the darkest night.
A conduit of power, it ignites the spark of life,
Guiding us towards a future, free from pain and strife.

In technology's embrace, it breathes innovation's fire,

A conductor of change, our dreams it does inspire.
From humble beginnings, it blazes forth with might,
Pushing boundaries, opening new realms of light.

Palladium, a guardian, a shining beacon of hope,
In every atom, resilience and strength elope.
With each discovery, with each life it mends,
It reminds us of the power that science extends.

So let us celebrate this element divine,
A symbol of progress, a star that will forever shine.
Palladium, our guide, on this journey we embark,
Leading us towards a future, illuminated and stark.

TWENTY-EIGHT

PRECIOUS ELEMENT

In medicine's realm, Palladium gleams,
A healer's touch, a poet's dreams.
With grace and might, it takes its stride,
A precious element, by nature's side.
    Within its core, a power resides,
A gift to humanity, it confides.
A catalyst of life, a potent elixir,
Palladium's touch, a divine mixture.
    Through ancient scrolls and whispered lore,
Its secrets unravel, forevermore.
In labs and clinics, its presence known,
A beacon of hope, when all seems blown.
    From ailments fierce, it brings relief,
A soothing balm, a gentle belief.

It mends the broken, with tender care,
Palladium's grace, beyond compare.

A metal of wonders, it does ignite,
The engines of progress, with fiery light.
In the heart of machines, it sparks the flame,
Innovation's dance, forever its aim.

With each passing day, Palladium reigns,
A symbol of strength, that never wanes.
In technology's embrace, it finds its place,
A guardian of change, a boundless embrace.

Oh, Palladium, you endure and persist,
In medicine's realm, and where progress exists.
A shimmering testament, to human ambition,
Your legacy shines, a divine composition.

TWENTY-NINE

CATALYST OF HEALTH

In the realm of medicine, a gleaming light,
Palladium, a healer, brings hope in the fight.
A catalyst of health, a guardian of life,
It soothes the ailments, eases the strife.
    Within its core, a power untold,
A metal divine, a treasure to behold.
With grace and poise, it mends the weak,
Restoring harmony, emboldening the meek.
    From ancient times to modern days,
Palladium's touch, a beacon that stays.
It whispers of solace, a promise of cure,
A testament to resilience, forever secure.
    In labs and clinics, its virtues unfold,
A symbol of progress, a story untold.

From broken bones to hearts that ache,
Palladium's magic, no ailment can break.
So let us celebrate this element rare,
Its presence in medicine, beyond compare.
With every discovery, a future untied,
Palladium, our guide, forever by our side.
Oh, Palladium, we sing your praise,
For the miracles you bring, in countless ways.
Through your strength and endurance, we find our way,
To a world of healing, brighter every day.

THIRTY

SECRETS RESIDE

In the realm of medicine's embrace,
A catalyst of healing grace,
Palladium, the element divine,
A symbol of progress, so sublime.

Within its structure, secrets reside,
Unleashing powers, none can hide,
A key to mending, a cure to find,
Palladium's touch, a gentle bind.

From lab to clinic, it takes its flight,
Igniting hope, dispelling plight,
Innovations born from its gleaming core,
Technology's triumph, forevermore.

A guardian of health, it stands tall,
A beacon of light amidst the pall,

Palladium, the healer's guide,
A steadfast force, by our side.
    Its noble presence, ever bright,
Guiding us through the darkest night,
A promise of futures yet untold,
Palladium's story, forever bold.
    Through its essence, progress takes flight,
Unveiling vistas, bathed in light,
With every breakthrough, hope is restored,
Palladium's legacy, forever adored.
    Oh, precious element, so rare and true,
We sing your praises, anew.
In medicine's realm, you bring relief,
Palladium, our eternal belief.

THIRTY-ONE

PALLADIUM'S PRESENCE

In realms of innovation, where wonders take flight,
There lies a noble element, gleaming and bright.
Palladium, the conductor of dreams and desire,
Ignites the embers of progress, higher and higher.

A catalyst of change, it weaves a potent spell,
Unleashing the potential within, as stories tell.
Its presence, a marvel, in technologies untold,
From circuits to catalysts, its worth manifold.

A guardian of secrets, it embraces the spark,
Unleashing in laboratories, a vibrant embark.
With strength and resilience, it forges ahead,
Pushing the boundaries of what lies ahead.

In medicine's realm, it offers its hand,
A healer, a comforter, a solace so grand.

With gentle touch, it eases the pain,
Bringing relief to hearts, again and again.
    Enduring and steadfast, it withstands the test,
A symbol of hope, when life is distressed.
Palladium, a beacon, shining through the night,
Guiding humanity towards a future so bright.
    So let us rejoice in its shimmering grace,
And marvel at its power, as we embrace.
For in Palladium's presence, we find the key,
To unlock the mysteries of what can be.

THIRTY-TWO

CHANGE AND HOPE

In realms where change and progress intertwine,
There shines a metal, precious and divine.
Palladium, conductor of the new,
A symbol of what innovation can do.

With silver hue, it captures every ray,
Reflecting light and guiding our way.
A catalyst, igniting transformation's fire,
Unleashing potential, taking us higher.

In circuits and wires, it finds its home,
Conducting energy, never to roam.
A conductor of change, in technology's hand,
Palladium lights our path, in this digital land.

But beyond the wires, its magic extends,
To medicine's realm, where healing transcends.

A guardian of health, a soothing balm,
Relieving pain, bringing serenity's calm.
 Its presence brings relief, both gentle and strong,
A beacon of hope when everything feels wrong.
In hospitals and labs, its power unfolds,
Bringing solace to hearts, as life's story unfolds.
 Resilient and enduring, it stands the test of time,
Guiding humanity towards a future sublime.
Palladium, element of miracles and might,
A reminder that progress is within our sight.
 So let us honor this metal, shining and rare,
For it holds the promise of a world that we share.
Palladium, conductor of change and hope,
In your presence, we find the strength to cope.

THIRTY-THREE

BRIGHT AND NEW

In the realm of technology, bright and bold,
A marvel of metal, Palladium we behold.
With its lustrous glow, a sight to admire,
It fuels our inventions, igniting the fire.

From circuits to screens, it weaves its charm,
A conductor of progress, a catalyst so warm.
In the heart of innovation, it takes its place,
Empowering our dreams, with elegance and grace.

Oh, Palladium, you're the key to our dreams,
Unleashing the power of endless streams.
In our gadgets and devices, you shine so bright,
Guiding us forward, with your radiant light.

But beyond the realm of circuits and wire,
Palladium's worth goes even higher.

In the realm of healing, it plays its part,
A savior of lives, a mender of heart.
    In the depths of medicine, it lends a hand,
A guardian of health, a beacon so grand.
With its healing touch, it eases our pain,
Restoring our bodies, like a gentle rain.
    Oh, Palladium, you're the healer we seek,
With your touch, ailments turn weak.
In the realm of wellness, you stand tall,
A symbol of hope, for one and all.
    So let us cherish this element divine,
From technology's realm to the healing shrine.
Palladium, you're a gift, both rare and true,
Guiding us towards a future bright and new.

THIRTY-FOUR

HARBINGER OF CHANGE

In the realm of healing, a shimmering light,
Palladium emerges, a beacon so bright.
A catalyst of science, a noble metal,
With powers untold, in its very mettle.

    Within the lab, it dances with grace,
Unveiling secrets, a celestial embrace.
A healer it becomes, a doctor unseen,
A remedy for ailments, a mystical dream.

    Its touch brings relief, like a gentle breeze,
Calming the storm, putting minds at ease.
Through its wisdom, diseases are tamed,
A guiding force, in a world inflamed.

    Oh, Palladium, guardian of health,
A symbol of hope, a source of great wealth.

With every discovery, a new chapter unfolds,
Unveiling the mysteries that nature holds.
    Enduring and strong, a sentinel of light,
Palladium's presence, a beacon in the night.
A catalyst for progress, a harbinger of change,
In its elemental form, a power so strange.
    So let us celebrate this noble metal,
A symbol of strength, resilience, and mettle.
For in its essence, we find the key,
To heal, to prosper, to set our spirits free.

THIRTY-FIVE

PALLADIUM, THE HEALER

In the realm where science and healing entwine,
A noble element, Palladium, doth shine.
A guardian of health, a precious alloy,
Its touch brings relief, a comforting joy.

    Within the lab, where remedies are sought,
Palladium whispers, its secrets are taught.
A catalyst of life, it aids in the fight,
Against ailments that plague, with all of their might.

    Oh, Palladium, thou art a beacon of light,
Guiding humanity through the darkest of night.
A symbol of hope, of endurance so strong,
In thy presence, we find solace, where we belong.

    Through the trials we face, and the battles we fight,
Palladium's grace, like a soothing light.

Its electrons dance, in a mystical sway,
Bringing healing and comfort, day after day.

So let us celebrate this element divine,
Palladium, the healer, whose touch is benign.
With gratitude we honor, its power and might,
A testament to science, and nature's delight.

ABOUT THE AUTHOR

Walter the Educator is one of the pseudonyms for Walter Anderson. Formally educated in Chemistry, Business, and Education, he is an educator, an author, a diverse entrepreneur, and he is the son of a disabled war veteran. "Walter the Educator" shares his time between educating and creating. He holds interests and owns several creative projects that entertain, enlighten, enhance, and educate, hoping to inspire and motivate you.

> Follow, find new works, and stay up to date
> with Walter the Educator™
> at WaltertheEducator.com

www.ingramcontent.com/pod-product-compliance
Lightning Source LLC
LaVergne TN
LVHW010602070526
838199LV00063BA/5048